How to Speak to a Robot

Mastering Ai Communication
with Clarity and Creativity

Mark Hallink

SPRINGWELL
PUBLISHING

How To Speak To A Robot

© 2025, Mark Hallink

Published By: Springwell Publishing, Newberry, Florida

First Edition 1-2-3-4-5-6-7-8-9-10

All Rights Reserved

Printed in the United States

ISBN (Hardback): 978-1-9717559-9-1
ISBN (Paperback): 979-8-9922405-6-6
ISBN (eBook): 979-8-9922405-5-9
ISBN (Audiobook): 979-8-9923032-3-0

Library of Congress Control Number: 2025911479

Also by Mark Halink

Pumpkin's Pennies

The Money Game

Quiet Money

CONTENTS

INTRODUCTION

Welcome to *How to Speak to a Robot,* the ultimate guide to navigating the ever-evolving world of artificial intelligence. This isn't your typical tech manual or a deep dive into coding. Instead, it's a conversational and practical guide to unlocking the true potential of AI in your life, whether you're a seasoned tech enthusiast or someone who still thinks 'the cloud' is just where you store vacation photos.

At its heart, this book is about communication—the art of interacting with AI effectively, efficiently, and intelligently. It's about treating AI not just as a tool, but as a partner, one that can help you achieve your goals, solve problems, and even spark creativity. But like any relationship, success depends on how well you communicate.

The chapters in this book are designed to take you on a journey, starting with the basics and building toward mastery. Each chapter focuses on a critical aspect of AI interaction, from asking the right questions to refining responses, setting clear goals, and even understanding the ethical and practical implications of working with robots. Along the way, we'll explore real-world examples, share tips, and bust some common myths about what AI can and can't do.

WHY THIS BOOK MATTERS

We live in a world where AI powers everything from our search engines to our smart home devices. Yet, many people feel frustrated or intimidated by the idea of interacting with these systems. The problem isn't the technology—it's how we approach it. AI, for all its complexity, is ultimately a mirror. The quality of its output depends on the clarity of our input. This book is here to help you provide that clarity. In this way you can provide useful input so the system can give you back useful output.

Whether you're:

• Crafting a business plan using AI,
• Exploring creative writing with a robot co-author,
• Troubleshooting technical problems, or
• Simply trying to get better directions from your GPS,

This book will teach you how to make AI work for you, not the other way around.

WHAT YOU'LL LEARN

Here's a sneak peek at what the chapters cover:

1. **Understanding the Basics**: What AI is, what it isn't, and how it works behind the scenes.
2. **The Art of Asking Questions**: Why specificity is key and how to phrase queries for maximum impact.
3. **Refining Responses**: How to guide AI when its answers miss the mark.
4. **AI as a Creative Partner**: Unlocking your imagination with the help of a robot assistant.
5. **The Rules of Engagement**: Ethical considerations and setting boundaries with AI.
6. **Teaching AI Your Goals**: The importance of clear objectives and how to align AI with your ambitions.
7. **Building an AI Strategy**: How to use AI to enhance productivity and decision-making.
8. **The Role of Intuition**: Balancing human judgment with machine intelligence.

9. **The Future of Communication**: What's next for AI and how to prepare for it.

10 **Speaking Robot Fluently**: A practical guide to lifelong learning in the age of AI.

WHO THIS BOOK IS FOR?

This book is for anyone who interacts with AI, whether you're a business leader, creative professional, student, or tech enthusiast. It's for those who want to stay ahead of the curve in a world where AI is becoming as ubiquitous as electricity. And most importantly, it's for anyone who believes learning to communicate better—with humans or robots—is a skill worth mastering.

So, grab a cup of coffee (or tea, if that's your thing), settle in, and get ready to embark on a fascinating journey. By the time you reach the final chapter, you'll be fluent in the language of robots and ready to harness AI's power like never before.

MEET CHATGPT – YOUR NEW AI SIDEKICK

Before we dive deeper into mastering AI communication, let me introduce you to one of the most versatile and accessible tools in the AI world—ChatGPT.

If you're wondering, "What exactly is ChatGPT?", think of it as the ultimate multitasking robot brain, designed to help you with practically anything you throw at it:

– Writing: Draft emails, write essays, create marketing copy, or brainstorm stories.

– Learning: Simplify complex topics, explain concepts, or help with study plans.

– Problem-Solving: Debug code, suggest ideas, or outline solutions.

- Conversation: Engage in creative discussions, ask for advice, or just chat about your day.

The key here isn't that ChatGPT replaces your ideas—it enhances them. It's like having a Swiss Army knife for creativity and productivity right at your fingertips.

There's a Bot for That

As AI has grown, so has the range of specialized bots. Whatever problem you're facing, odds are, "There's a bot for that." Let's meet a few:

1. Productivity Bots:

- Notion AI: Helps summarize notes, generate ideas, and improve organization within the popular Notion app.
- Otter.ai: Transcribes meetings and interviews in real time, so you never miss a detail.

2. Customer Support Bots:

- Tools like Intercom and Zendesk AI handle repetitive customer service queries, offering quick solutions to everyday problems.

3. Creative Bots:

- MidJourney and DALL·E generate artwork and visuals based on prompts.
- Runway ML assists with AI-driven video editing and visual effects.

4. Finance Bots:
  – Cleo helps track spending, save money, and make budgets more manageable through a conversational, humorous chat interface.
  – Bloomberg AI assists traders by analyzing data to make quick decisions.

5. Health Bots:
  – Ada Health helps users analyze symptoms and suggest possible causes.
  – Replika is designed as a chatbot to help with emotional well-being and loneliness.

ChatGPT and Beyond: Why It Matters

The magic of ChatGPT—and AI bots in general—is that they're not just tools for tech enthusiasts. They're for everyone: students, writers, artists, business leaders, and even folks just looking to make life a little easier.

For example:
  – Struggling to plan a family vacation? There's a bot for that.
  – Need help learning a new language? There's a bot for that.
  – Want a recipe based on the random ingredients in your fridge? You guessed it—there's a bot for that.

These AI companions might not replace human creativity or judgment, but they'll certainly get you to the finish line faster—and often with surprising results.

Understanding the Robot Brain

THE MYSTERIOUS MACHINE

Have you ever asked AI for help, only to feel like you're talking to someone who doesn't quite speak your language? The truth is, you are. AI is like an incredibly well-read, but overly literal alien—it has access to vast amounts of knowledge, but zero common sense.

Before we dive into *how* to talk to AI, let's start with *why* it works the way it does.

HOW AI THINKS (SPOILER: IT DOESN'T)

Artificial intelligence isn't sentient. It doesn't 'know' anything in the way humans do. Instead, it predicts what words or phrases are most likely to come next based on patterns it's learned. It's like a highly advanced autocomplete, but instead of just finishing your sentences, it can write essays, debug code, or even plan a party—if you ask the right way.

Here's the gist:

1. **AI Takes Your Input Literally**: It processes exactly what you say, not what you mean.

2. **AI Sees Patterns**: It uses training data to predict the best response, but it doesn't 'understand' your intent unless you explain it clearly.
3. **Context Matters**: The more context you give, the better the results.

WHAT AI LOVES: CLARITY

Imagine you're giving directions to someone blindfolded. You wouldn't just say, "Go over there." You'd explain the steps: "Take three steps forward, turn left, and walk until you feel a wall." AI works the same way. The clearer and more detailed your input, the better the output.

The Problem with Vague Questions

Here's an example of how vague input can backfire:

- **Question**: "Tell me about dogs."
- **Response**: A generic summary about dogs, likely not what you wanted.

Now compare it to a specific question:

- **Question**: "Tell me about the history of golden retrievers as guide dogs."
- **Response**: A focused, detailed answer about golden retrievers and their training for guiding.

The takeaway? Be specific.

AI'S STRENGTHS AND WEAKNESSES

AI excels at:

- Analyzing large amounts of data.
- Answering well-defined questions.
- Providing creative ideas when prompted.

But it struggles with:

- **Ambiguity**: "Write something cool" is too vague.
- **Human Emotions**: AI doesn't feel, so it can't empathize.
- **Nuance**: If your question has layers of meaning, you may need to guide it through each layer.

YOUR ROLE: THE TRANSLATOR

Think of yourself as the bridge between human creativity and machine logic. Your job is to translate your goals into clear, actionable instructions that AI can process. The better you get at this, the more productive your interactions will be.

Quick Tips for Better Results

1. **Start Simple**: Begin with a straightforward question or request.
2. **Provide Context**: Explain the goal or background of your request.
3. **Iterate**: If the first response isn't perfect, refine your question.

Your First Exercise

Let's practice! Try asking the following:

1. **A vague question**: "What's the weather like?"
2. **A specific question**: "What's the weather in Miami, Florida, today?"

Compare the results. You'll see how a little clarity makes all the difference.

Up Next: Crafting the Perfect Question

Now that you understand how AI processes information, the next chapter will dive into crafting the kinds of questions that lead to great answers. Spoiler: It's easier than you think.

Crafting the Perfect Question

THE SECRET SAUCE

Have you ever asked a vague question and been frustrated when the answer was equally unhelpful? Talking to AI is no different. The quality of your results depends on the quality of your input. A good question is like a well-built bridge—it connects your intention with the AI's capabilities.

In this chapter, we'll explore how to craft questions that lead to clear, precise, and useful answers.

WHAT MAKES A QUESTION 'PERFECT'?

A perfect question has three key elements:

1. **Clarity**: It's easy to understand and free from ambiguity.
2. **Specificity**: It defines exactly what you're looking for.
3. **Context**: It provides enough background for the AI to understand your request.

Let's break these down.

1. **Be clear**. AI is literal. It doesn't assume, guess, or read between the lines. If your question is unclear, the response will be, too.

Example:
- **Vague**: "Tell me about space."
- **Clear**: "Explain the benefits of space exploration for renewable energy development."

By removing ambiguity, you guide the AI toward a focused answer.

2. **Be Specific**. The more specific your question, the better the response. Vague requests often yield generic or irrelevant answers.

Example:
- **Vague**: "How do computers work?"
- **Specific**: "Explain how quantum computers differ from classical computers in processing information."

Specificity helps AI zero in on the exact information you need.

3. **Provide Context**. AI doesn't know what's in your head—it only knows what you tell it. Adding context helps the AI tailor its response.

Example:
- **Without Context**: "Write a summary."
- **With Context**: "Summarize this article for a 10-year-old audience."

Adding details like the purpose, audience, or desired tone improves the relevance of the output.

THE POWER OF EXAMPLES

Sometimes, showing is better than telling. Including examples in your prompt can clarify what you're looking for.

Example:

- **Without Examples**: "Write a product description."
- **With Examples**: "Write a product description similar to this: "This sleek, lightweight laptop is perfect for professionals on the go."

Examples help AI align its response with your expectations.

AVOIDING COMMON PITFALLS

Here are some mistakes to avoid when crafting questions:

1. **Overloading**: Asking for too much at once can confuse the AI.

   Example: Instead of "Explain quantum mechanics, write a summary of Einstein's contributions, and recommend beginner books," break it into smaller tasks.

2. **Assuming Knowledge**: Don't expect AI to infer missing details.

   Example: Instead of "Help me fix this," describe the problem: "My computer won't boot and displays a blue screen."

3. **Ignoring Iteration**: Rarely will the first answer be perfect. Ask follow-ups to refine the response.

ITERATIVE QUESTIONING: A REAL-WORLD EXAMPLE

Let's say you're planning a trip to Paris. Here's how iterative questioning works:

1. **Initial Question**: "What are the best things to do in Paris?"
   - **Response**: A generic list of tourist attractions.
2. **Follow-Up**: "Focus on activities for art lovers."
   - **Response**: A list of museums and galleries.
3. **Refinement**: "Include hidden gems off the beaten path."
   - **Response**: A tailored itinerary with lesser-known art spots.

By refining your question with each response, you guide AI to deliver exactly what you need.

THE GOLDEN RULE: THINK LIKE A LIBRARIAN

A good librarian doesn't just point you to a book—they ask clarifying questions to understand your needs. Treat AI the same way. Be specific, give context, and don't be afraid to dig deeper.

YOUR TURN: PRACTICE MAKES PERFECT

Try crafting a perfect question for the following scenarios:

1. You're writing a blog post about sustainable living.
2. You need help brainstorming a birthday gift for your 12-year-old niece.
3. You're researching the history of electric vehicles.

Compare your results when you add clarity, specificity, and context. You'll see how small tweaks make a big difference.

UP NEXT: THE PROBLEM WITH PERFECTION

In Chapter 3, we'll explore why perfection isn't always the goal. Sometimes, the best results come from experimentation, flexibility, and knowing when to let the AI surprise you.

The Problem
with Perfection

WHY PERFECTION ISN'T ALWAYS THE GOAL

When working with AI, it's easy to fall into the trap of chasing the 'perfect' response. You might spend hours refining prompts, tweaking phrases, or trying to predict exactly how the AI will react. But here's the truth: perfection is overrated.

In this chapter, we'll explore why aiming for flexibility and creativity often leads to better outcomes than obsessing over a single perfect answer.

AI'S STRENGTH ISN'T PERFECTION

AI thrives in the realm of possibilities. It doesn't think like a human, and it's not limited by human biases or constraints. This makes it a powerful brainstorming tool, capable of generating ideas or solutions that might never occur to you.

Example:
- You ask for a new product idea in sustainable fashion.
- AI suggests a line of biodegradable shoes made from seaweed—a concept you'd never considered.

Instead of forcing AI to match your exact vision, let it explore the edges of what's possible. You might be surprised by the results.

THE ROLE OF EXPERIMENTATION

Think of your interaction with AI as an experiment. The goal isn't to get it 'right' on the first try—it's to explore, iterate, and refine until you find something useful.

Example of Iteration:
1. **Initial Prompt**: "Write a tagline for an eco-friendly brand."
   Response: Sustainability Made Simple.
2. **Follow-Up**: "Make it more playful."
   Response: Green Is the New Black.
3. **Follow-Up**: "Focus on the brand's innovative technology."
   Response: Smart Tech for a Greener Tomorrow.

Each step moves closer to what you need, but the process allows room for creativity.

LETTING AI SURPRISE YOU

Sometimes, the best answers come from letting go of control. Give AI a bit of freedom, and it might exceed your expectations.

Example:
   Prompt: "Write a short poem about space exploration."
   Response: A whimsical verse about astronauts planting gardens on Mars.

You didn't ask for a poem with that specific theme, but the result is unique and engaging. By embracing unpredictability, you open the door to innovation.

WHEN 'GOOD ENOUGH' IS BETTER THAN PERFECT

In many situations, a 'good enough' answer is all you need to move forward. Spending too much time polishing a response can lead to diminishing returns.

Example:
- You're drafting an email and need a professional closing line.
- Instead of testing ten variations, go with, "Looking forward to your response," and move on.

AI is a tool to save time, not add unnecessary complexity. Learn to recognize when an answer is sufficient for the task at hand.

BALANCING PERFECTION AND PRACTICALITY

While perfection isn't always necessary, there are times when precision is important—such as legal documents, technical specifications, or anything involving high stakes. In these cases:

1. Use AI as a starting point, not the final authority.
2. Double-check facts and details.
3. Edit with human judgment to ensure accuracy.

For everything else, allow some room for flexibility.

BUILDING A FEEDBACK LOOP

One way to find balance is by creating a feedback loop with AI. Here's how it works:

1. Ask for a response.
2. Analyze what works and what doesn't.
3. Refine your prompt based on that analysis.
4. Repeat until satisfied.

This iterative approach helps you strike the right balance between exploration and precision.

PRACTICE EXERCISE: EMBRACE IMPERFECTION

Try this exercise:

1. Ask AI to brainstorm five ideas for a creative project (e.g., a new novel, app, or business).
2. Evaluate the responses. Which ones spark new thoughts, even if they're not exactly what you expected?
3. Use one of the ideas as a starting point and refine it further.

Notice how the process evolves as you build on the AI's suggestions.

THE TAKEAWAY

Perfection is an illusion. The real power of AI lies in its ability to generate options, spark creativity, and adapt to your needs. Embrace the imperfections, and you'll discover new possibilities you hadn't even considered.

UP NEXT: THE POWER OF ITERATION

In Chapter 4, we'll dive deeper into the iterative process, exploring how to refine responses step by step and turn rough drafts into polished gems.

The Power of Iteration

WHY ITERATION MATTERS

N o one writes a perfect sentence, paints a masterpiece, or solves a complex problem on the first try. Great results come from the process of iteration—trying, refining, and improving until you get it right. Working with AI is no different. The more you treat AI as a collaborative partner in an ongoing process, the better your outcomes will be.

In this chapter, we'll explore how to master the art of iteration and turn rough drafts into polished results.

START BROAD, THEN NARROW

The first step in iteration is to start with a general request and refine from there. Think of the initial response as a rough draft—a starting point on which to build.

Example: Writing a Marketing Tagline

1. **Initial Prompt**: "Create a tagline for a new fitness app."
   Response: Your Health, Your Way.
2. **Refinement Prompt**: "Make it more energetic."
   Response: Move Faster. Feel Stronger.

3. **Final Prompt**: "Focus on innovation and personalization."
Response: Revolutionize Your Fitness Journey.

Each iteration brings you closer to a tagline that aligns with your vision.

ANALYZE WHAT WORKS

After each response, take a moment to evaluate. What parts of the output align with your needs? What's missing? Identifying the strengths and weaknesses of each iteration helps you refine your request.

Example: Planning a Speech

1. **Prompt**: "Write an opening for a speech about climate change."
Response: Climate change is the defining issue of our time.
2. **Analysis**: The opening is impactful, but too general. You want a personal anecdote.
3. **Refinement Prompt**: "Add a personal story about seeing the effects of climate change."
Response: I remember standing on the shore as the waves eroded a beach I had visited as a child.

This iterative approach transforms a generic response into something meaningful and tailored.

THE ROLE OF FEEDBACK

Iteration works best when you provide clear, constructive feedback to the AI. Think of feedback as a way to guide the AI toward your goals.

Example of Feedback:
1. **Original Prompt**: "Summarize this article."
   Response: This article discusses renewable energy.
2. **Feedback Prompt**: "Include specific examples of renewable energy sources mentioned in the article."
3. **Refined Response**: This article highlights solar, wind, and hydroelectric power as key renewable energy sources.

By giving feedback, you clarify your needs and improve the quality of the response.

WHEN TO STOP ITERATING

While iteration is powerful, it's also possible to overdo it. At some point, the returns diminish, and the effort of refining outweighs the benefits. Learn to recognize when the response is 'good enough' for your needs.

Signs It's Time to Stop:
1. The response meets your requirements.
2. Further changes are minor or cosmetic.
3. The result aligns with your goals and feels complete.

Pro Tip: Don't let perfectionism get in the way of progress. Sometimes, 'done' is better than 'perfect.'

ITERATING FOR COMPLEX PROJECTS

For large or multi-step tasks, break the project into smaller pieces and iterate on each one individually. This makes the process more manageable and ensures each part gets the attention it needs.

Example: Writing a Business Proposal
Step 1: Outline the sections of the proposal.
- Iterate until you have a clear structure.

Step 2: Draft each section (e.g., executive summary, market analysis).
- Refine one section at a time.

Step 3: Combine the sections and polish the final document.
- Iterate on the overall flow and tone.

PRACTICE EXERCISE: STEP-BY-STEP ITERATION

Choose a task and use iteration to refine it:

1. **Task**: Plan a family vacation.
2. **Initial Prompt**: "Suggest vacation ideas for a family of four."
3. **Follow-Up Prompts**:
   - "Focus on destinations within a 5-hour flight."
   - "Include activities for children under 10."
   - "Provide cost estimates for a 7-day trip."

Compare the results at each step. Notice how the final plan becomes more detailed and actionable.

THE TAKEAWAY

Iteration isn't just a technique—it's a mindset. By embracing the process of refining and improving, you unlock the full potential of AI and turn rough ideas into polished results. The best collaborations happen when you allow room for growth and discovery.

UP NEXT: COLLABORATION IN ACTION

In Chapter 5, we'll dive into real-world examples of how iteration and collaboration come together to tackle challenges, solve problems, and create something extraordinary.

When Robots Get It Wrong

(And Why It's Not the End of the World)

I f you've spent even five minutes talking to a robot, you've probably encountered one of those moments where the response makes you pause, tilt your head, and mutter, "What on Earth were you thinking?" Spoiler: The robot wasn't thinking—it was calculating, and sometimes, those calculations miss the mark.

This chapter is about what happens when AI gets it hilariously, frustratingly, or bafflingly wrong—and how to turn those moments into opportunities (or at least a good laugh).

WHY ROBOTS FAIL (AND IT'S USUALLY NOT THEIR FAULT)

Robots, as clever as they are, have limitations. Here's why they sometimes drop the ball:

1. Ambiguous Prompts

If you ask, "Write about cheese," don't be surprised when you get a random mashup of Gouda facts, grilled cheese recipes, and a paragraph on why lactose intolerance is a global issue. AI needs context to deliver.

2. Training Data Blind Spots

AI is trained on vast amounts of data, but it's not omniscient. It doesn't know what it hasn't been taught. If your prompt dives into obscure medieval knitting techniques, you might get a very creative (but totally inaccurate) response.

3. Overconfidence in Output

AI doesn't know when it's wrong. If you ask it to explain quantum physics, it might confidently give you an answer—but that doesn't mean the answer is correct. It's not lying; it just doesn't know any better.

4. The User Factor

Let's face it—sometimes, the problem isn't the robot. If your prompts are vague, confusing, or unrealistic, the results will be, too.

FAMOUS AI FAILS (AND WHAT WE CAN LEARN FROM THEM)

Here are some examples of AI falling short and the lessons they teach us:

1. The Mistranslation Mishap

Prompt: "Translate 'I'm full' into Japanese." Output: "私は完全です" (which means "I am perfect").
Lesson: Language is tricky. Always double-check translations with a native speaker or credible source.

2. The Creative Stretch

Prompt: "Write a limerick about entropy."

Output: There once was a process quite messy, Where order turned into spaghetti... (And then it just stops).

Lesson: Robots may start strong, but struggle with complex creative tasks.

3. The Literal Interpretation

Prompt: "Write a story about a hero who saves the world."

Output: Once upon a time, a hero saved the world. The end.

Lesson: Be specific! If you want drama, details, and depth, say so.

HOW TO FIX A FAILING ROBOT

When AI doesn't deliver, don't panic. Here's how to troubleshoot:

1. Refine Your Prompt

Vague Prompt: "Tell me about history."

Improved Prompt: "Summarize the key events of World War II in 200 words, focusing on Europe."

2. Break It Down

Complex Prompt: "Write an essay on climate change, including its causes, effects, and solutions."

Better Approach:

Step 1: "Explain the causes of climate change."

Step 2: "Now explain the effects."

Step 3: "Suggest solutions."

3. Adjust Your Expectations

Sometimes, the robot just can't do what you're asking. If your prompt is overly ambitious or requires human-level creativity, consider rethinking it.

4. Laugh It Off

Remember: Robots are tools, not geniuses. If they mess up, take it as a chance to laugh and learn. Who doesn't love a good AI blooper?

WHEN TO TRUST AI (AND WHEN NOT TO)

Here's a quick guide to knowing when to rely on your robot sidekick:

- **Trust It For:**
  - Generating ideas (e.g., brainstorming blog topics).
  - Summarizing straightforward information.
  - Automating repetitive tasks (e.g., writing email templates).
- **Double-Check For:**
  - Factual accuracy. (Google is still your friend.)
  - Sensitive topics or cultural nuances.
  - Situations requiring emotional intelligence or judgment.
- **Avoid Asking:**
  - "What should I do with my life?" (That's a question for a therapist, not a robot.)
  - "Write me a bestseller" (AI can help, but the heart of the story has to come from you).

TURNING FAILURES INTO WINS

Every AI fail is an opportunity to refine your approach. Here's how to make the most of those moments:

1. Learn the Limits

The more you experiment, the better you'll understand what AI can and can't do. Use failures as a way to map the boundaries.

2. Improve Your Communication

If the robot didn't get it, maybe the prompt wasn't clear enough. Think of it as a conversation: What could you rephrase or add?

3. Embrace the Unexpected

Sometimes, the mistakes are more interesting than what you originally wanted. That spaghetti entropy poem? It might inspire something better.

PRO TIP: FAIL FORWARD

Think of talking to AI like training a puppy. Mistakes are inevitable, but they're part of the process. Over time, you'll both get better at understanding each other (and neither will chew up your shoes).

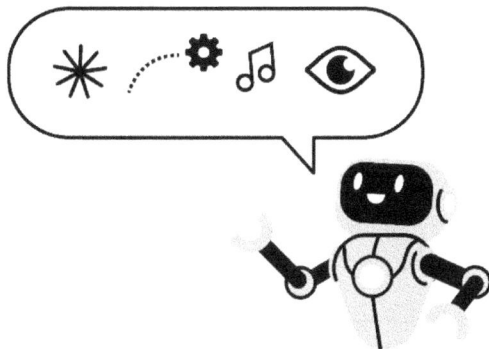

FINAL THOUGHTS

AI isn't perfect—and that's okay. In fact, some of the most delightful and surprising moments come from when it gets things wrong. So the next time your robot suggests you pair peanut butter with pickles, don't get frustrated. Instead, take a step back, refine your prompt, and maybe have a little laugh. After all, learning is half the fun.

Next up: Let's explore how to use AI to level up your creativity, from storytelling to brainstorming big ideas. Ready to unlock your inner genius? Let's do this!

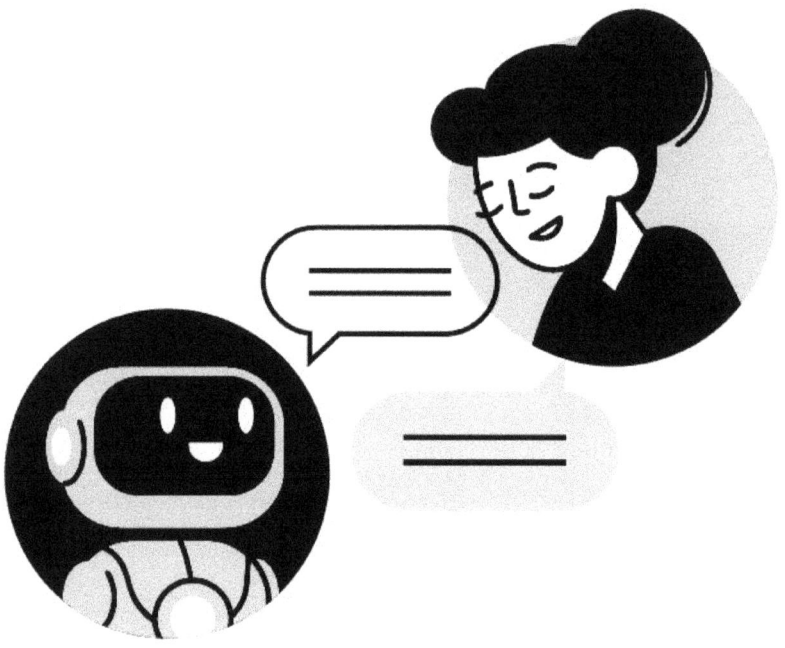

Teaching AI to Understand Your Goals

So, you've figured out how to ask questions, get recommendations, and maybe even draft the perfect email with your AI assistant. But here's a secret: the real power of AI lies not just in what you ask, but in how well it understands your goals. Think of it as the difference between a friend who knows you're into fishing and one who knows you're on a mission to catch the biggest marlin this summer.

WHY GOALS MATTER IN AI COMMUNICATION

AI thrives on specificity. When you're clear about your objectives, the AI can leverage its vast capabilities to help you achieve them. Without clear goals, your interactions might feel like wandering in a fog. You'll get somewhere, but it might not be where you intended to go.

For example:
- Instead of asking, "What are the best books on productivity?" try, "What books on productivity can help me manage my time better as a freelance writer?"

This subtle shift gives the AI a roadmap, narrowing its focus and improving the relevance of its response. (Which brings to mind, WHERE does AI get their information exactly, technically, HOW does that happen, and HOW is it then delivered to us? I know this might sound like a silly question, but these are the things we are thinking about here as we read this so we KNOW other people like us will have them too, particularly people who have zero technical knowhow but are interested in learning more about AI.)

STEP 1: DEFINE THE GOAL

Before you even start typing, think about what success looks like in your interaction. Are you looking for a detailed plan? A quick answer? Inspiration? Knowing this upfront will sharpen your queries.

Take this example:
- Vague goal: "I want to get fit."
- Defined goal: "I want a six-week workout plan that fits into my 30-minute daily routine at home with no equipment."

The clearer your goal, the better the AI can guide you.

STEP 2: CONTEXT IS KING

AI doesn't (yet) have the ability to read minds. If your query lacks context, it will make assumptions—and not always the right ones. Provide background information upfront to save time and avoid frustration.

For instance:
- Instead of, "What's the best way to budget?"
- Try, "I make $50,000 a year and want to save for a down payment on a house in five years. What's the best budgeting method for me?"

The AI now has a much better foundation to deliver tailored advice.

STEP 3: ITERATE AND REFINE

Even with clear goals, your first attempt might not yield the perfect response. That's okay. Iteration is part of the process. AI learns best when you guide it with feedback. Use phrases like:

- "That's helpful, but can you give me examples specific to [my situation]?"
- "Can you simplify this?"
- "Can you explain this from a beginner's perspective?»

Each tweak brings AI closer to understanding your unique needs.

PRO TIP: COLLABORATE, DON'T COMMAND

Think of your interaction with AI as a partnership rather than a transaction. Share your thought process, ask for alternative perspectives, and encourage AI to think creatively. For example:

- "I need help planning a trip. Here's what I've thought of so far—can you suggest improvements?"
- "What are the pros and cons of this approach? Is there something I might be overlooking?"

This collaborative approach can yield surprising insights.

COMMON MISTAKES WHEN SETTING GOALS FOR AI

1. **Being Too Broad**: Asking, "How do I get rich?" is not going to get you far. Instead, break it down: "What are some investment strategies for someone with $5,000 to start?"

2. **Neglecting Constraints**: If you have time, budget, or skill limitations, mention them. AI can't account for restrictions you don't specify.

3. **Not Following Up**: Treating AI's first answer as gospel can lead to missed opportunities. Probe deeper.

FINAL THOUGHTS

When you teach AI to understand your goals, you're not just getting better answers—you're building a smarter assistant. Over time, this relationship can grow, as the AI learns your preferences and priorities. Remember, the more effort you put into clarifying your goals, the more effective your AI will be.

And who knows? By the end of this book, you might even teach the robots to think like you—just without the caffeine addiction.

7

Collaborating with AI

THE PARTNER YOU DIDN'T KNOW YOU NEEDED

By now, you've learned how to ask the right questions and get the most from AI. But what if we took things a step further? What if we treated AI not just as a tool but as a partner? After all, collaboration isn't about issuing commands—it's about working together toward a shared goal.

In this chapter, we'll explore how to approach AI as a collaborator, leverage its strengths, and even adapt when things don't go as planned.

THE ART OF COLLABORATION

Collaboration is all about complementing each other's strengths. AI excels at:

1. **Data Crunching**: It can analyze vast amounts of information in seconds.
2. **Generating Ideas**: Need 50 brainstorming ideas? AI's got you.
3. **Repetitive Tasks**: From formatting text to creating schedules, it's happy to do the grunt work.

But there are things only *you* can bring to the table:

1. **Judgment**: You know what makes sense in the real world; AI doesn't.
2. **Creativity**: While AI can help generate ideas, it doesn't have a vision.
3. **Nuance**: You understand subtlety, humor, and emotion in ways AI simply can't.

Together, you form an unbeatable team—as long as you know how to play to each other's strengths.

ASSIGNING ROLES TO AI

AI is like an actor waiting for direction. If you don't give it a clear role, it'll improvise (and not always well). Here's how to assign roles for better results:

- **The Expert**: "You are an economist. Explain the impact of inflation on small businesses."
- **The Assistant**: "Help me summarize this report for a presentation."
- **The Creator**: "Write a short story about a time-traveling historian."
- **The Critic**: "Review this essay and suggest improvements."

By giving AI a specific role, you guide its responses and ensure it aligns with your needs.

WHEN COLLABORATION GETS MESSY

Sometimes, working with AI feels like herding cats. Here's how to handle common challenges:

1. The AI Goes Off-Track

- Problem: You ask for help with an essay, and AI starts writing a grocery list.
- Solution: Reframe the request. Provide more context or clarify the goal.

2. The Output Feels Generic

Problem: The response lacks depth or originality.

Solution: Ask for a deeper analysis or specify what kind of answer you need. For example, "Explain with examples" or "Focus on historical events."

3. The Response Is Wrong

Problem: AI confidently delivers an incorrect answer.

Solution: Double-check the facts. AI isn't perfect, and neither is its data.

ITERATIVE COLLABORATION: THE KEY TO SUCCESS

Collaboration with AI is rarely a one-and-done process. Think of it like sculpting—each response is a rough draft you refine over time. Here's how to iterate effectively:

1. **Start Broad**: Begin with a general question or task.
2. **Refine**: Analyze the response and ask follow-up questions to narrow the focus.
3. **Repeat: Keep** refining until you're satisfied with the results.

Example:
- **First Request**: "Summarize climate change."
- **Follow-Up**: "Focus on the economic impact of climate change in developing countries."
- **Final Request**: "Provide examples of policies that have worked to mitigate these impacts."

BUILDING TRUST WITH YOUR AI PARTNER

Trust might sound like a strange word to use with AI, but it's important. The more you experiment, the better you'll understand its capabilities and limitations. Over time, you'll learn when to trust its answers and when to dig deeper.

Case Study: A Real Collaboration

Let's say you're writing a business plan for a startup. Here's how you might collaborate with AI:

1. **Define the Goal**: "Help me create a business plan for a mobile app that tracks personal carbon footprints."
2. **Assign a Role**: "Act as a sustainability expert."
3. **Iterate**: Use AI's initial outline, then refine sections like market analysis or financial projections by asking for specific details.
4. **Final Touches**: Add your personal insights, ensuring the plan reflects your vision and values.

THE HUMAN-AI TEAM

The best collaborations happen when both parties contribute their strengths. You bring the creativity, judgment, and vision; AI brings the data, speed, and structure. Together, you can tackle problems faster, explore new ideas, and achieve goals that might have seemed out of reach.

NEXT STEPS

In the next chapter, we'll look at advanced techniques for pushing your AI collaborations even further. From combining tools to creating complex workflows, you'll learn how to get the absolute most from your robot partner.

8

Advanced Techniques for AI Collaboration

LEVELING UP YOUR AI GAME

By now, you've mastered the basics of communicating with AI and treating it as a collaborator. But what if you could push the boundaries even further? Advanced AI techniques can help you tackle complex problems, create more nuanced content, and even automate parts of your workflow.

This chapter will explore strategies for maximizing your productivity and creativity with AI. We'll dive into combining tools, crafting layered prompts, and creating workflows that integrate AI into your daily life.

LAYERED PROMPTS: BUILDING COMPLEXITY

Sometimes, you need more than a single answer. Layered prompts allow you to guide AI through a multi-step process, ensuring detailed and thorough results.

Here's how it works:

1. **Break Down the Task**: Identify the smaller steps within a larger goal.
2. **Guide the Process**: Ask questions or make requests in sequential order.
3. **Synthesize Results**: Combine the outputs into a cohesive whole.

Example: Writing a White Paper

Step 1: "Summarize the latest trends in renewable energy."

Step 2: "List the challenges and opportunities associated with solar energy."

Step 3: "Draft an introduction to a white paper using these insights."

By layering prompts, you create a structure that keeps AI focused and on track.

COMBINING TOOLS FOR MAXIMUM IMPACT

AI is powerful on its own, but pairing it with other tools can take your productivity to new heights. Here are some examples:

1. **Writing and Editing**: Use AI for drafting, then polish your work with grammar tools like Grammarly or Hemingway Editor.
2. **Data Analysis**: Pair AI with spreadsheet software for detailed calculations and visualizations.
3. **Creative Projects**: Combine AI-generated content with design tools like Canva or Adobe Creative Suite.

Pro Tip: Think of AI as the engine and other tools as the steering wheel. Together, they help you reach your destination faster and more effectively.

CREATING CUSTOM WORKFLOWS

If you frequently perform similar tasks, consider designing a workflow that integrates AI seamlessly. Here's an example for project management:

1. **Brainstorming**: Use AI to generate ideas or identify potential risks.
2. **Planning**: Ask AI to outline steps, create timelines, or assign responsibilities.
3. **Execution**: Collaborate with your team, using AI to streamline communication and documentation.
4. **Review**: Have AI analyze performance metrics or draft summaries for stakeholders.

Workflows like this save time and reduce mental load, freeing you to focus on high-value activities.

PROMPT ENGINEERING: FINE-TUNING FOR PRECISION

Prompt engineering is the art of crafting highly specific inputs to achieve optimal results. Here are some techniques:

1. **Role Assignment:** "Act as a historian specializing in medieval Europe."
2. **Constraints**: "Summarize this article in 100 words or fewer."
3. **Examples**: "Rewrite this paragraph in the style of a New York Times op-ed."

Experimenting with different prompt structures helps you discover what works best for your goals.

SOLVING COMPLEX PROBLEMS

AI can be a valuable partner for tackling big challenges, but you'll need to guide it carefully. Here's an approach for solving complex problems:

1. **Define the Problem**: Be as specific as possible about the issue you're addressing.
2. **Gather Perspectives**: Ask AI to provide multiple viewpoints or solutions.
3. **Evaluate Trade-offs**: Discuss the pros and cons of each option.
4. **Refine the Solution**: Work iteratively to arrive at the best outcome.

Example: Designing a Sustainable Building

Prompt 1: "List innovative materials for eco-friendly construction."

Prompt 2: "What are the cost implications of using these materials?"

Prompt 3: "Develop a preliminary proposal for a zero-energy office building."

ADVANCED COLLABORATION CASE STUDY

Imagine you're leading a product development team. Here's how advanced AI techniques could support your efforts:

1. **Research**: Use AI to analyze market trends and consumer preferences.
2. **Design**: Generate mockups or concept descriptions for brainstorming sessions.
3. **Testing**: Simulate potential user feedback using AI-generated personas.

4. **Launch**: Draft press releases, marketing strategies, and customer outreach plans.

With AI as your co-pilot, you can accelerate every stage of the process.

BALANCING AUTOMATION AND HUMAN INSIGHT

While AI can handle many tasks efficiently, it's essential to remember its limitations. Automation is a tool—not a replacement for human judgment, creativity, or empathy. Use AI to amplify your capabilities, but never lose sight of your unique contributions.

NEXT STEPS

Chapter 9 will take us even deeper into the possibilities of human-AI collaboration. We'll explore real-world applications in various fields—education, business, healthcare, and more—and see how people are using AI to solve problems and create value.

Real-World Applications of AI

TURNING THEORY INTO PRACTICE

We've explored how to interact with AI, treat it as a collaborator, and even integrate it into workflows. But how does this look in the real world? In this chapter, we'll examine how AI is transforming industries, solving problems, and unlocking opportunities.

Whether you're an entrepreneur, educator, or hobbyist, these examples will inspire you to think creatively about what's possible with AI.

AI IN EDUCATION: TEACHING THE NEXT GENERATION

Education is one of the most promising frontiers for AI. Teachers and students alike are using AI to make learning more personalized, accessible, and engaging.

Examples:

1. **Personalized Learning Plans**: AI analyzes students' strengths and weaknesses to recommend tailored study

paths.

2. **Automated Grading**: Tools like AI-powered essay graders save teachers hours of work.
3. **Interactive Tutoring**: Virtual assistants provide instant feedback and explanations in subjects ranging from math to language arts.

Impact: By reducing administrative tasks, educators can focus on teaching, while students receive the individualized attention they need to thrive.

AI IN BUSINESS: DRIVING INNOVATION

From startups to global corporations, businesses are leveraging AI to optimize operations and enhance decision-making.

Examples:

1. **Market Analysis**: AI processes massive datasets to identify trends and opportunities faster than any human team could.
2. **Customer Service**: Chatbots handle common inquiries, freeing up human agents for more complex issues.
3. **Product Design**: AI tools like generative design help engineers create innovative solutions by simulating countless possibilities.

Impact: Businesses gain a competitive edge by making data-driven decisions and delivering more value to their customers.

AI IN HEALTHCARE: SAVING LIVES

In healthcare, AI is revolutionizing diagnostics, treatment planning, and patient care.

Examples:
1. **Early Detection**: AI-powered imaging tools identify diseases like cancer or heart conditions at earlier stages than traditional methods.
2. **Drug Discovery**: Machine learning accelerates the development of new medications by analyzing molecular data.
3. **Telemedicine**: Virtual assistants support doctors by triaging patients and summarizing medical records.

Impact: Faster, more accurate diagnoses and treatments improve outcomes and save lives.

AI IN CREATIVE FIELDS: UNLOCKING IMAGINATION

AI is also a tool for creativity, empowering artists, writers, and musicians to explore new possibilities.

Examples:
1. **Writing Assistance**: AI helps authors brainstorm ideas, refine drafts, and even co-write stories.
2. **Music Composition**: Tools like AI-powered synthesizers generate melodies and harmonies based on user input.
3. **Visual Arts**: Artists use AI to create digital paintings, animations, and generative designs.

Impact: AI expands the creative toolkit, allowing individuals to bring ambitious projects to life.

AI IN EVERYDAY LIFE: MAKING THINGS EASIER

AI isn't just for professionals—it's making everyday tasks simpler and more efficient.

Examples:
1. **Smart Assistants**: Devices like Alexa and Siri manage schedules, answer questions, and control smart home devices.
2. **Language Translation**: AI breaks down language barriers with real-time translation apps.
3. **Shopping Recommendations**: E-commerce platforms suggest products based on browsing history and preferences.

Impact: AI saves time, reduces friction, and enhances convenience for millions of people.

WHAT CAN YOU DO WITH AI?

No matter your field or interests, AI has something to offer. Here's how you can start incorporating AI into your own work:
1. **Experiment**: Try AI tools in your area of interest—whether it's writing, coding, or managing tasks.
2. **Learn from Others**: Look for case studies or examples of how others are using AI successfully.
3. **Stay Curious**: The AI landscape is evolving rapidly, so keep exploring new tools and techniques.

BALANCING POTENTIAL AND RESPONSIBILITY

While AI offers incredible opportunities, it also comes with challenges. Ethical considerations, data privacy, and the risk of over-reliance are all issues to keep in mind. The key is to use AI thoughtfully, ensuring it supports, rather than replaces, human ingenuity.

NEXT STEPS

In the final chapter, we'll reflect on the future of human-AI collaboration. We'll discuss how to stay ahead of the curve, embrace continuous learning, and approach AI with optimism and responsibility.

The Future of Human-AI Collaboration

THE JOURNEY SO FAR

Congratulations! You've made it to the end of this guide, and by now you've learned how to communicate effectively with AI, treat it as a partner, and apply it in real-world scenarios. But this is only the beginning. Artificial intelligence isn't just a tool—it's a revolution in how we work, think, and create.

In this final chapter, we'll look ahead to what's next. How will AI continue to shape our lives? What challenges lie ahead? And most importantly, how can we prepare for a future where human-AI collaboration is the norm?

THE EVOLVING ROLE OF AI

AI is advancing at a pace that few could have imagined even a decade ago. Here are some areas where we can expect breakthroughs in the near future:

1. **Personalized Everything**: From healthcare to education, AI will tailor solutions to individual needs with unprecedented precision.

2. **Advanced Problem-Solving**: AI will tackle complex challenges like climate change, global health crises, and sustainable development.
3. **Creative Synergy**: Artists, writers, and musicians will continue to push the boundaries of creativity with AI as their co-creator.

But as AI evolves, so too must we evolve. The ability to adapt, learn, and collaborate will be more important than ever.

CHALLENGES ON THE HORIZON

The future of AI isn't without its challenges. Here are some key issues we must navigate:

1. **Ethics and Bias**: AI systems are only as unbiased as the data they're trained on. Ensuring fairness and accountability will be critical.
2. **Job Displacement**: Automation will reshape industries, and we'll need to find ways to retrain and support workers in transition.
3. **Data Privacy**: As AI relies on ever-larger datasets, protecting individual privacy will become an even greater priority.
4. **Over-Reliance**: The convenience of AI can lead to complacency. Maintaining human oversight and critical thinking will be essential. (Many people will be wondering about the continued discussion if AI should be regulated. Any thoughts to share on this?)

THE HUMAN ADVANTAGE

While AI is powerful, it has limitations that ensure humans will always have a role to play. Here's what sets us apart:

- **Judgment**: The ability to make nuanced decisions based on values and ethics.
- **Empathy**: Understanding emotions and building meaningful connections.
- **Creativity**: The spark of innovation that comes from seeing the world in new ways.

By combining our strengths with AI's capabilities, we can achieve more together than either could alone.

PREPARING FOR THE FUTURE

How can you stay ahead in a world increasingly shaped by AI? Here are some practical tips:

1. **Embrace Lifelong Learning**: The pace of change means staying curious and adaptable is more important than ever.
2. **Stay Informed**: Follow developments in AI to understand its capabilities and implications.
3. **Focus on What AI Can't Do**: Develop skills in areas where human expertise remains unmatched, like interpersonal communication, strategic thinking, and creativity.

A VISION FOR COLLABORATION

Imagine a world where AI is seamlessly integrated into every aspect of life:

- Farmers use AI to optimize crop yields and combat food shortages.
- Doctors rely on AI to diagnose and treat patients with pinpoint accuracy.

- Students around the world access personalized education, tailored to their unique learning styles.

This isn't science-fiction—it's the future we're building together.

THE FINAL TAKEAWAY

As you move forward, remember that AI is a tool—a partner that amplifies your abilities but doesn't replace them. The key to thriving in an AI-powered world is not fear or resistance, but curiosity and collaboration.

So go ahead: experiment, learn, and embrace the possibilities. The future of human-AI collaboration is bright, and you're ready to be a part of it.

THANK YOU FOR READING

Thank you for taking this journey with me. I hope this book has not only answered your questions about AI, but also inspired you to ask new ones. After all, the best way to speak to a robot is to approach it with an open mind—and a little bit of humor.

Now, let's see what you and your AI partner can create.

ACKNOWLEDGMENTS

First and foremost, I extend my deepest gratitude to my wonderful and patient wife, Alice, whose undying support and love have been my anchor and inspiration throughout the creation of "The Money Game."

Alice, your unwavering encouragement and the countless ways you assist me daily have been pivotal in this endeavor. This journey of exploring and elucidating complex financial concepts would not have been as enriching or achievable without you by my side. I must also express my appreciation for the array of online tools that have significantly enhanced my research and writing process.

These digital resources have been indispensable, not only for drafting and refining this manuscript, but also for connecting with talented individuals such as the voice actors who brought the audio version of this book to life, and the editors and proof-readers who polished the text to perfection. The availability and accessibility of such tools have truly transformed the creative landscape for writers everywhere. Lastly, I wish to acknowledge the assistance provided by ChatGPT.

Its role in helping me refine ideas, clarify financial principles, and even navigate the complexities of narrative structure has been invaluable. Its help has been a vital part of this project, and for that, I am incredibly grateful.

To everyone who has contributed, whether mentioned here by name or remembered in spirit, thank you for helping transform "The Money Game" from a concept into a reality that I hope will enlighten and inspire many.

We'd Love to Hear Your Thoughts!

If you have enjoyed reading this book or have any feedback, we'd greatly appreciate your review on Amazon. As a new author your feedback is incredibly important and very much appreciated. Thank you for your support and for helping to spread the word!

Please scan the QR code below to leave a review.

ABOUT THE AUTHOR

Mark Hallink and his wife, Alice, are "trailblazers" in the world of Cowboy Mounted Shooting and dedicated advocates for the sport. As the first man and woman in Canada to achieve Men's Level Six (M6) and Ladies' Level 6 (L6), respectively, the highest level in the discipline, Mark and Alice's accomplishments have inspired countless enthusiasts to pursue their own riding and shooting goals. Professionally, Mark was the founder and former president of Hallink RSB Inc., a leading innovator in the tooling and packaging industry. His company played a key role in advancing technology and design in many of the products found on your grocery store shelves, setting industry standards for quality and efficiency. After selling the company in 2020, Mark transitioned into retirement, where he now focuses on his passions: writing and spending time with the love of his life, Alice, and his favorite horse, Pumpkin. Mark studied at the University of Waterloo in Canada, earning an Honours Degree in Psychology. Mark's writing spans topics like financial literacy, where he shares his expertise in business and entrepreneurship, as well as fiction, where he explores imaginative stories inspired by his experiences and love for adventure. His work reflects his belief in perseverance, financial independence, and the joy of creative expression. Mark and Alice are retiring to North Florida, where they plan to embrace the region's natural beauty, and friendly people, and continue their equestrian pursuits, exploring new adventures around the world.

BOOKS BY MARK HALLINK

Pumpkin's Pennies

In *Pumpkin's Pennies*, young readers embark on a heartwarming journey with Pumpkin as she discovers the value of hard work, saving, spending wisely, and sharing. Set in the vibrant world of Springwell Farms, the story follows Pumpkin's determination to earn her first "horsey coins," save for a warm blanket and eventually use her financial knowledge to lead her community toward prosperity.

The Money Game

The Money Game is a compelling guide designed to transform your approach to personal finance. It unravels the hidden rules of the money game, combining insights from economics, psychology, and personal experience to teach you how to navigate financial challenges with wisdom and humor. Hallink offers practical strategies for saving, investing, and managing resources, emphasizing the importance of understanding the interplay between money, time, energy, and relationships. With a touch of humor and real-life examples, this book equips readers to achieve financial success by making smarter choices and understanding the subtleties of financial freedom.

Quiet Money

Quiet Money: Growing Wealth Quietly and Why It's Super Important is a comprehensive guide to building and maintaining wealth with a focus on living a fulfilling life, aligned with personal values rather than societal expectations. The book rejects the flashy, ostentatious displays of wealth prevalent on social media and emphasizes a more intentional, humble approach to financial success.

www.ingramcontent.com/pod-product-compliance
Lightning Source LLC
Chambersburg PA
CBHW040930210326
41597CB00030B/5248